iScience Readers

Environments:
Beetles in the Garden

by Emily Sohn and Barbara M. Linde

Chief Content Consultant
Edward Rock
Associate Executive Director, National Science Teachers Association

NORWOODHOUSE PRESS
Chicago, Illinois

Norwood House Press
PO Box 316598
Chicago, IL 60631

For information regarding Norwood House Press, please visit our website at
www.norwoodhousepress.com or call 866-565-2900.

Special thanks to: Amanda Jones, Amy Karasick, Alanna Mertens, Terrence Young, Jr.

Editors: Michelle Parsons, Diane Hinckley
Designer: Daniel M. Greene
Production Management: Victory Productions, Inc.

Paperback ISBN: 978-1-60357-289-7

Printed in Heshan City, Guangdong, China.
190P—082011.

CONTENTS

iScience Puzzle   6

Discover Activity   8

What Is an Environment?   10

What Are the Different Types of Biomes?   14

Sun, Wind, Rain, and Snow   17

The Roles of Plants and Animals   18

What Causes Stress In an Ecosystem?   23

Science at Work   28

Nature's Recycling System   29

How Do Ecosystems Change Naturally Over Time?   32

Connecting to History   35

The Human Factor   38

Solve the iScience Puzzle   42

Beyond the Puzzle   44

Glossary   45

Further Reading/Additional Notes   46

Index   48

Note to Caregivers:

Throughout this book, many questions are posed to the reader. Some are open-ended and ask what the reader thinks. Discuss these questions with your child and guide him or her in thinking through the possible answers and outcomes. There are also questions posed which have a specific answer. Encourage your child to read through the text to determine the correct answer. Most importantly, encourage answers grounded in reality while also allowing imaginations to soar. Information to help support you as you share the book with your child is provided in the back in the **Additional Notes** section.

Words that are **bolded** are defined in the glossary in the back of the book.

How Will You Save the Roses?

Gardens may burst with plants like flowers, herbs, fruits, and vegetables. But plants can't grow just anywhere. They need certain things, such as sunlight, water, and help from people. Just like gardens, every **environment** around the world has its own look and its own needs. In this book, you'll learn about all sorts of environments and what makes them tick. You'll also solve a puzzle about a rose garden under attack.

Beetles in the Garden

You run to your garden. You've been away for a week. You are eager to smell the flowers. But when you get there, you're shocked. Many of the roses in your garden are dead! The leaves are full of holes. On the plants, you see hard-shelled bugs with copper-colored wings. Japanese beetles have invaded your garden! What is the best way to save your garden from these insects?

Japanese beetle

Solution 1: Allow a Predator In

Moles, skunks, bats, some wasps, and some birds eat Japanese beetles.

Solution 2: Use a Pesticide

Pesticides are chemical poisons. You spray them on the beetles, on the plants, or on both.

Solution 3: Use a Trap

Beetle traps emit odors that attract beetles. The insects get stuck in a bag that is part of the trap. Then, you can destroy the bag.

Solution 4: Do Nothing

Maybe the roses will get better by themselves.

Solution 5: Get Rid of the Roses

Instead, plant flowers that Japanese beetles do not eat.

Some robins eat adult Japanese beetles.

Before you decide what to do, ask yourself these questions:
- How will adding something to the garden change it?
- Will your choice hurt anything or anyone?
- Will the method attract more beetles?
- Is the solution safe?
- Do you have a better idea than what's listed here?

Understanding Stability

If you removed the beams that support your house, your house would fall down. In the same way, beetles have upset the structure of your rose garden. The environment is no longer stable. When something is stable, it is firm and steady. A stable environment can put up with some amount of change. For example, a stable house can hold up against most storms. But when an environment becomes unstable, changes to it can make it fall apart.

You can demonstrate the idea of **stability** by building a house of cards.

Here's another way to think about stability: Grab a deck of cards. Balance them to build a structure. Now experiment with the shape and height of the structure. What makes the house more likely or less likely to topple over?

Now, build several card houses. Each should look different. Place a penny on the top level, or second story, of one structure. Put a penny on the bottom level, or first story, of another. If a structure fell down, did the whole thing break? Or did just part of it fall down? How could you change its shape so that the structure would be more stable?

Think about your rose garden. How might each possible solution affect the garden's ability to stay strong?

What Is an Environment?

Like children, each garden is different. It holds certain types of flowers, insects, and rodents. It has its own kind of dirt. It experiences unique weather patterns, called climate. Together, these details add up to make a place what it is. The whole area is called the environment. Mountains, rivers, and playgrounds are other kinds of environments.

Urban environments are crowded with people and buildings. However, many have parks where people can relax.

Fire can change an environment quickly. This picture shows forest fire damage at Big Cypress National Preserve in Florida.

There are lots of ways to describe a place. Physical details include land, water, and climate. Environments are also full of biological details. These include plants and animals, such as flies and dandelions. All living things are part of an environment.

Environments change all the time. Some change quickly. Others change slowly. Changes might be big or small. Over time, these changes add up. A place can look very different from one decade to the next.

Look around you. What kinds of plants and animals live in your environment? What does the land look like? How is your environment different from other places? What might make your environment more stable or less stable?

Forest habitats support many plants and animals.

How to Describe a Place

Environments are filled with nooks and crannies that get filled with life. Places where plants and animals live are called **habitats.** Your rose garden is a habitat for some creatures. Meadows and even city parks are habitats, too.

Most living things need habitats that contain food and water. Animals need fresh air to breathe. Plants need certain nutrients in the dirt.

Together, plants, animals, and their habitats make up an **ecosystem.** Each **species** in an ecosystem does its own thing. One species may climb trees and eat fruit. Another might hunt for small rodents.

What happens when a habitat does not meet the needs of a **population**? The population might move to a better habitat. Or it might start dying off. Look back at the rose-garden puzzle to solution 4: Do Nothing. What do you think would happen to the environment if you let the beetles eat the rose bushes? What would happen to the roses? What would happen to the beetles?

What living things would you find in this ecosystem?

When you're inside with a book, nature can feel far away. But people are part of ecosystems, too. Look at the land behind your school. Walk along a street near your home. Even if you live in a city, you'll see plants, trees, and animals.

Go outside and inspect the first ecosystem you find. What lives there? What kind of habitats do you see?

You might notice more than one ecosystem around you. A group of ecosystems is called a biome. Biomes describe a whole region. They include climate, soil, plants, and animals. All of these details are different from one place to the next. A cactus garden is an ecosystem within a biome called a desert. These are specific ways of talking about the environment around us.

Think about the plants and animals in your rose garden. Now, imagine you could look at the garden from high in the sky. What part of the world is your rose garden in? What are the weather and seasons like? What kinds of dirt and rocks lie on the ground?

If you were to circle the globe, you'd see biomes up high and down low. There are biomes with lots of rocks or mud. There are biomes with lots of trees.

Ponds, lakes, streams, rivers, and wetlands are all part of freshwater biomes. The water there does not have much salt in it. Birds and fish live in these environments. Plants grow near the water's edge. Some grow roots in the water.

This coral reef in the Red Sea is part of a marine biome.

A marine biome has salty water. Oceans, coral reefs, and **estuaries** are parts of marine biomes. Fish, lobsters, sponges, and other animals live there. Seaweed is a major source of food for marine animals.

Above water, mountain biomes are full of variety. From the bottom of a mountain to its top, you can see changes in weather, soil, and plant life. There is less oxygen at higher **elevations.** The animals and plants that live on mountaintops are often pretty tough. They may have to deal with cold temperatures. Landscapes are rugged. And there is not a lot of oxygen to breathe at high **altitudes.**

This grassland is in Ireland.

Grasslands are another type of land biome. In these environments, grass beats out trees and shrubs. Animals, such as zebra and bison, eat the grasses.

Forests are bustling biomes, too. One way to tell the difference between types of forests is to look at the kinds of plants that grow there. Tall trees grow in some forests because there are lots of nutrients in the soil. Evergreens, ferns, and mosses grow in rainforests. These are very wet environments. Oak, maple, and elm trees live in forests that have four seasons. These trees shed their leaves to survive the winter. Pine, fir, and spruce trees have tough needlelike leaves that survive in cold northern forests.

The weather and plants in a biome help the animals survive. How do you think some biomes meet the needs of people?

Desert biomes are dry places.

But wait, there's more! Other land biomes include deserts and tundra. A desert biome gets less than 20 inches (50 cm) of rain a year. In many deserts, days are hot and nights are cold. Desert animals are usually small and active at night. The soil is mostly sandy or made of rough gravel. Cacti are found in some deserts. These plants have thick stems that are good at holding on tight to water.

Tundra are the coldest biomes on Earth. A tundra gets about 10 inches (25 cm) of rain or snow a year. The soil is frozen into a layer called permafrost. Plants grow there. But they are short and have shallow roots. Animals in the tundra often **hibernate** or **migrate** to survive the long winter.

As you can see, each biome is a sum of its plants, animals, soil, and climate. What do you think would happen to the stability of a biome if the soil changed?

Sun, Wind, Rain, and Snow

Sunny. Rainy. Cloudy. The weather might affect what you do and how you feel. Over years and years, weather conditions start to make patterns. Together, the average of these conditions is called the climate. Weather describes what it feels like right now. Climate describes the big picture.

Flooding from rain can cause stress in a forest.

An ecosystem's climate determines which plants and animals can live there. Cacti thrive in hot, dry desert climates. Polar bears live in cold climates. What would happen to a polar bear that was moved to the desert? How would a cactus fare in the frigid Arctic?

Many climates experience severe weather. Snowstorms, hurricanes, and tornadoes are a few examples. These events can both help and hurt ecosystems. A flood can enrich the soil next to a river. But it might also destroy plant life and animal habitats.

How would you describe the weather right now where you live? How would you describe your climate? How does the climate where you live affect your life in both good and bad ways?

Everything that lives has a job. Plants are called **producers.** They take energy from sunlight and turn it into a form that life can use. This process is called **photosynthesis;** it allows plants to make their own food. Animals are called **consumers.** They can't make their own food. Instead, they have to eat what they can find in the environment.

Green plants get energy from the Sun using photosynthesis.

Different animals eat different things. Herbivores eat mostly plants. Carnivores eat other animals. Animals that eat both plants and animals are called omnivores. In your rose garden, Japanese beetles consume rose leaves.

Think about where you live. What plants and animals live there? What do they eat? How do they depend on each other for food and shelter?

The porcini mushroom, a fungus, is a decomposer.

Eating the Leftovers

Decomposers have the dirtiest jobs. They eat dead plants and animals. It may sound gross, but it's important work. Decomposers turn complex chemicals into simple ones. The process puts out nutrients, such as carbon and nitrogen. These nutrients go into the air, the soil, and the water. They allow the cycle of life to continue. Fungi, earthworms, and bacteria are decomposers.

The award for "sneakiest creatures" goes to **scavengers.** They eat dead animals that they did not kill themselves. Some beetles, termites, and flies are scavengers. So are cockroaches, vultures, rats, opossums, and catfish.

What kinds of decomposers live in your rose garden? How do decomposers make the garden more stable or less stable? What would happen to the garden if there were no decomposers?

The coyote is the predator and the bird is its prey.

Catch or Be Caught

Nature is like a big game of tag. Some animals chase. Others are chased. **Predators** are animals that hunt other animals for food. They're the "It" in a round of tag. The animals that are hunted are called **prey.** Some animals are both predator and prey.

This chart shows some predator-prey relationships.

Predator	Prey
human	birds, fish, other animals
coyote	birds, rabbits, rodents
bat	mosquitoes, beetles, moths
eagle	fish, snakes, rodents
lion	zebras, antelope, giraffes
spider	flies, ants, mosquitoes, and other insects
praying mantis	beetles, crickets, grasshoppers
frog	insects
raccoon	frogs, mice, worms

Skunks eat the wormlike young of Japanese beetles.

Predator or prey, every animal needs to eat. Many animals are predators sometimes and prey at other times. Look at the chart on the previous page. Can you find an animal that is both predator and prey? When do you think it does the chasing? When is it chased?

Look at solution 1 of the rose-garden puzzle. Suppose you let a mole, skunk, or bat into your garden. All of these animals eat beetles. Is the new animal a predator or prey?

Predators often eat more than one type of prey. Before you welcome a predator into your garden, take a look around. What other animals live nearby? How might a new predator affect them?

Do you think it is a good idea to bring in a predator to eat the Japanese beetles?

Who Eats Whom?

One way to keep track of all the action in nature is to draw a picture. A food chain shows who eats whom. Look at the food chain below. In your garden, Japanese beetles eat roses. Skunks eat Japanese beetles. And coyotes eat skunks.

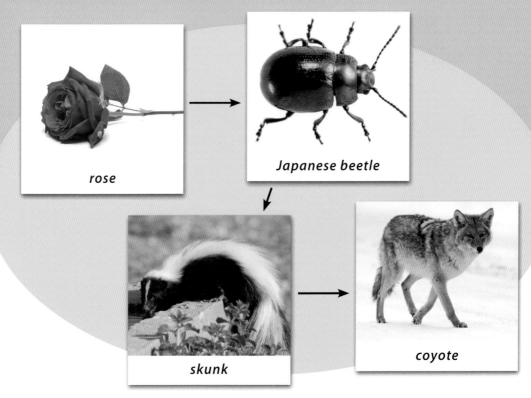

rose

Japanese beetle

skunk

coyote

Each connection is called a link. Most food chains have only a few links. Why do you think they aren't longer? What do you think would happen to the food chain if you removed all the Japanese beetles?

Sometimes, food chains cross other food chains. The result is a food web. Like spider webs, food webs branch in many directions. For example, coyotes don't eat just skunks. They also eat rats. What do you think rats eat? How could you add rats and other creatures to the diagram above? What would the food web look like?

What Causes Stress in an Ecosystem?

An ecosystem is like your house of cards. Placed carefully, the cards balance on each other. But with one strong push or gust of wind, they topple over. A stressor is something that causes an ecosystem to become less stable. Diseases, storms, and pollution cause stress. So do fires, climate change, and shifts in temperature.

Strong winds and rain put stress on a tropical ecosystem.

After going through stress, an ecosystem may find its balance again. That balance is called equilibrium. Other ecosystems cannot repair the damage. Instead, they change. In these cases, a new ecosystem will replace the old one.

Think about your rose garden. What would happen if you let the Japanese beetles stay? Would they put stress on the ecosystem? Do you think it is a good idea to pick solution 4: Do Nothing? Why or why not?

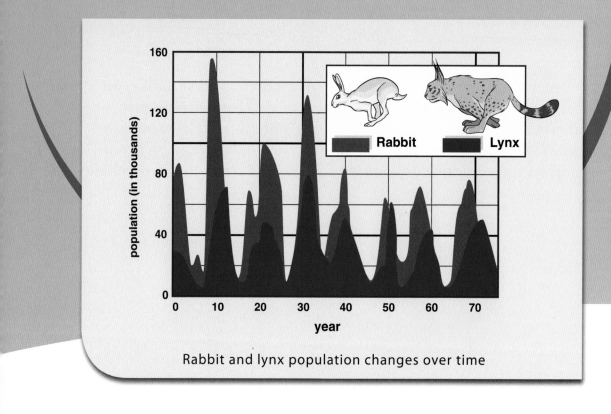

Rabbit and lynx population changes over time

Not Too Big, Not Too Small

Ecosystems are always changing. When times are good, populations grow. But they can't grow forever. After a while, a large population of animals would run out of food or space. Often, populations grow and shrink, over and over again. A population shouldn't become too big or too small. If it does, the ecosystem might not be stable anymore. Think about your garden puzzle again. How do you think solution 4: Do Nothing relates to population size?

This chart shows how numbers of rabbits and lynxes changed in an area over time. The graph covers a period of 70 years. How do the population sizes change in relation to each other? How do the two populations affect each other?

Some animals eat
Japanese beetles.

In the graph on the previous page, the lynx is the predator. The rabbit is the prey. Suppose the number of lynxes increases. There are only so many rabbits. So now, the lynxes must compete with other lynxes for their prey. What might happen to the lynx population after a while?

Now suppose the number of rabbits goes up. Do you think each lynx will eat as many rabbits as it can? If it does, what would happen to the number of rabbits? What would happen to the number of lynxes?

Think about the rose garden puzzle again. In solution 1, you allowed a predator into the garden. If you chose this solution, what would happen to the beetle population?

? Did You Know?

Japanese beetles first appeared in Japan. They were accidentally brought to the United States about 100 years ago. In Japan, beetle numbers stay low. That's because the weather there is cool. There aren't many good habitats for the beetles, and there are plenty of natural predators to control them. In the United States, conditions are much better for beetle populations to explode.

Mountains like these provide few resources for plants. The plants that live there have changed somehow in order to survive.

Running Low

Imagine that all of the animals in the zoo are locked in the giraffe's pen. Nobody is coming to help. All they have to eat and drink is what's left in the pen. As supplies run out, the animals will probably start to fight. In nature, the weakest members of the group sometimes die.

Over time, a population might develop **adaptations.** It learns to live with less. Or, if another habitat is available, the population may be able to move. That helps the animals. But the new habitat can suffer from stress. It is not used to the new animal. It may become less stable.

Honeybees are important to farmers.

Think about the beetles in your rose garden. Suppose you choose solution 5. You replace the roses with flowers that Japanese beetles do not eat. Suddenly, the beetles have much less food.

What do you think will happen to the beetle population? Will the population decrease in size? Will the beetles move to a different garden? Or will the beetles start eating something else in your garden? Do you think solution 5 is a good idea? Why or why not?

❓ Did You Know?

Honeybees help fruits and vegetables reproduce. Lots of honeybees have been dying lately. Scientists aren't sure why. As honeybee populations get smaller, what do you think will happen to the rest of the ecosystem?

Urban planners consider resources and people when planning communities.

SCIENCE AT WORK

Urban Planner

An urban planner decides what to do with land. City officials often talk to urban planners when they want to start new construction. There are a lot of things to think about. Planners need to meet the needs of people who live in the city. But they also need to think about the ecosystems in and around the city.

Urban planners study how new structures will affect the stability of a city. For example, a new building may add pollution. It could increase traffic. And it might make the area look better or worse.

Think about the neighborhood you live in. How would you change it to make it better? Consider the resources, such as drinking water and green space, which are already in your neighborhood.

Nature's Recycling System

Plants and animals don't have garbage cans. Instead, they use and reuse the stuff that keeps them alive. Water is one thing that gets recycled in nature. It is part of a cycle that repeats over and over.

This lake is part of the water cycle.

To start thinking about cycles, consider the water you use every day. Where does it come from? Where does it go? Does water you use end up in the ocean? Could it end up in a cloud? Do you ever reuse water? Do you ever waste water?

Water, Water Everywhere

You wouldn't drink the water that you flush down the toilet. But Earth recycles water all the time. The water cycle is the movement of water above, on, and below Earth's surface. The cycle has been going on for billions of years.

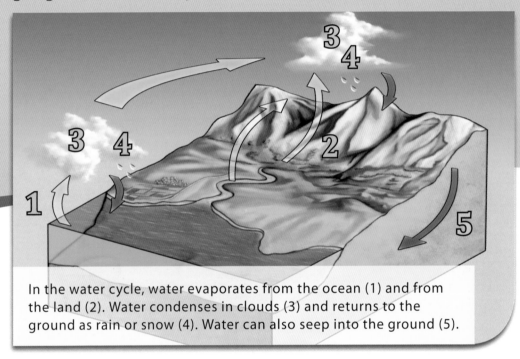

In the water cycle, water evaporates from the ocean (1) and from the land (2). Water condenses in clouds (3) and returns to the ground as rain or snow (4). Water can also seep into the ground (5).

Most of Earth's water is in the oceans. But the water doesn't always stay there. The Sun heats the water and it **evaporates,** or turns into water vapor. Vapor is a gas that rises into the air. As it rises, it gets colder. Eventually, water vapor **condenses,** or turns to liquid again. Clouds are made up of many drops of water. Rain or snow falls from clouds to Earth. The water from rain and snow builds up in oceans, lakes, and rivers. The water cycle starts all over again.

If you choose solution 2 for your rose garden, do you think the pesticides might end up in the ocean? How would they get there? What might happen if you used pesticides and then did not get rain for a long time?

sunlight

oxygen

carbon
dioxide

water

Photosynthesis is
one process in the
oxygen-carbon cycle.

More Cycles

Water isn't the only thing that cycles in the environment.
Another cycle involves two **elements** called oxygen and carbon.
You take part in the oxygen-carbon cycle every time you take a
breath. Like other animals, people breathe in oxygen from the air.
When we exhale, we breathe out carbon dioxide.

Plants do the opposite. They take in carbon dioxide. During
photosynthesis, they release oxygen into the atmosphere. Animals
use that oxygen to breathe. The cycle starts all over again. In the
process, plants and animals help each other survive.

Plants use their leaves to take in carbon dioxide. When the
beetles ate the leaves of your rose plants, why did the plants die?

No two days are alike. The same goes for months, years, and decades. Over time, the conditions of a place change. Reasons include storms, pests, diseases, fires, and floods. **Succession** describes how plant life changes after the land or climate changes.

For example, here's what happens after a forest fire. First, grass grows on the parched earth. Then shrubs grow, followed by trees. Over time, a dense forest develops.

Ferns cannot thrive without lots of moisture.

Picture an area where lots of ferns grow. Ferns need lots of water and some shade. What would happen if the climate changed, leading to less shade and less rain? What might happen to other plants that can live well in these new conditions?

How could a change in climate affect the stability of your rose garden? What might happen to the beetles?

Young, thin trees grow in this new forest.

The Life of a Forest

Imagine a forest. You watch as the forest is cut down to make way for a farm. After many years, the farmers abandon the land. The land is left on its own. What will happen next? Keep watching.

First, weeds, grasses, and flowers grow. Shrubs and some young trees appear next. These trees grow quickly in the sunlight. As the trees grow, they produce more shade. The sun-loving trees die out. Shade-loving trees replace them. These shade trees become a stable forest.

What do you think happened to animals in the area when the forest was first cut down? What will animals do when the forest grows back?

Volcanic eruptions can destroy plants and animals.

Saying Farewell to Earth

You've probably heard of dinosaurs and seen pictures in books. But no one has ever seen them in real life. That's because they are **extinct.** That means they don't live on Earth anymore. Many types of plants and animals have become extinct. Extinction can happen when members of a group aren't strong enough to fight for food or places to live. Disease, ice ages, storms, and volcanic eruptions can also destroy species. If a species cannot **adapt** to a new environment, it will die out.

Did you know that some types of plants and animals have been around for hundreds of millions of years? Ferns and horseshoe crabs are two examples. Can you think of other really old examples? Can you think of plants or animals that have gone extinct? Try looking on the Internet to find some.

Rachel Carson published *Silent Spring* in 1962.

Silent Spring by Rachel Carson (1907–1964)

Silent Spring is the name of a book written in 1962. The book describes a time when animals were dying. Songbirds were disappearing.

The author was Rachel Carson. She was a writer, marine biologist, and ecologist. She studied the environment. She also wrote articles about it. Her books about the ocean were popular.

Carson found out that the government allowed pesticides in natural areas. Farmers and businesses used them to kill insects. A pesticide called DDT was one of the most dangerous. Carson learned that DDT harmed people, animals, and the environment. She wrote *Silent Spring* to warn people about these dangers.

Salmon ate insects poisoned by DDT.

Silent Spring told scary stories about DDT and other pesticides. Sometimes, people used the chemicals to kill certain insects on balsam fir trees. But many other insects died, too. Salmon that swam in nearby rivers normally ate these insects. With their food source gone, the salmon died.

In the book, Carson warned people about breathing in pesticide sprays. She wrote about fruits and vegetables that were covered in chemicals. She also explained how pesticides get into groundwater. People could then get sick from drinking the water.

Carson worried about pesticides sprayed on farm vegetables.

The bad news kept coming. Carson wrote that pesticides stopped working after people used them for a while. Then, people needed to switch to stronger chemicals. These chemicals caused even more damage to living things and the environment.

Pesticide companies argued that Carson was wrong. They said that pesticides were more helpful than harmful. Many farmers, scientists, and some government officials agreed. Still, Carson's ideas drew lots of support. Citizens wrote letters to the government. In response, the government banned some pesticides.

Think back to solution 2 of the rose-garden puzzle. It suggested using pesticides to control Japanese beetles. What are the risks of using chemicals in your garden?

The Human Factor

Around the year 1600, sailors landed on a small island in the Indian Ocean. There, they found a type of bird called the dodo. The birds could not fly. The sailors cut down the island's fruit trees. After that, the birds didn't have food. Animals from the ships also ate the birds and their eggs. In just 80 years, the dodo bird disappeared from the island. They didn't live anywhere else. Dodos are just one example of how people affect environments.

The dodo stood 3 feet (1 meter) tall and weighed about 40 pounds (18 kilograms).

Whales are another example. Today, many types of whales are in danger. Some people are working to protect whales. But others still hunt whales.

Think about your rose garden. Why might you want to protect your roses and not protect the Japanese beetles? Should you consider protecting both? Do they both have a role in the garden ecosystem?

Today, there are laws in the United States to prevent this kind of harm to ecosystems.

How Does Pollution Affect the Stability of an Ecosystem?

Pollution is one of the biggest threats to ecosystems. Some urban areas have large factories that burn coal. The process sends smoke and soot into the air. Dirty air makes it hard to breathe. People and animals can get sick. On some days, it can be hard to see through the dirty air.

Acid rain is another kind of pollution. It damages plants, bodies of water, and even buildings. It forms when air pollution mixes with falling rain. Acid rain can make fish unsafe to eat.

Many gardeners add fertilizer to garden soil.

How Do Fertilizers Affect an Ecosystem?

Many farmers and gardeners use fertilizers to help their plants grow. Fertilizers are chemicals. When it rains, they can wash into nearby lakes and streams. There, the chemicals poison fish and other animals. The water becomes unsafe to drink or swim in. Fish become unsafe to eat.

Fertilizers also make some plants in the water grow and multiply faster. So many plants result in too many of some kinds of nutrients in the water. **Algae** that feed on the nutrients then multiply. As the algae die and are consumed by decomposing organisms, available oxygen in the water is used up. Fish then die.

Fertilizers cause algae to overgrow in lakes and rivers, suffocating plants and animals.

Streams or storms can carry water from one body of water to another. Then, the pollution spreads. Another problem is that fertilizers sink into the soil. There, they get into the roots of plants.

Fertilizers make your garden more stable because they enrich the soil and help the plants. What are the downsides? What are some other ways you might enrich the soil in your rose garden besides using fertilizers?

Now, let's return to the puzzle in your rose garden. You considered five possible solutions. The information in this book should help you make your decision. Here is a review of the solutions. All of the solutions have some upsides, called pros. They also have downsides, called cons.

Solution 1: Allow a Predator In

Pros: Allowing a predator does not involve chemicals.
Cons: The predator may eat helpful insects or plants that you like. It may tear up your garden. It could also scare your pets or friends.

Solution 2: Use a Pesticide

Pros: The chemical will probably kill the bugs.
Cons: The chemical may kill other insects, harm your pets, or poison the groundwater. After a while, the pesticide might not work on the beetles anymore.

Solution 3: Use a Trap

Pros: The trap may catch a lot of beetles without using chemicals.
Cons: All of the beetles in the neighborhood may come to your yard. They will come to find the sweet-smelling stuff in your traps. These pesky insects might eat your roses before they get to the trap.

Japanese beetle trap

Solution 4: Do Nothing

Pros: Nothing harmful is added to the environment.
Cons: The Japanese beetles will destroy your rose plants. You might have no flowers. Your garden will look ugly.

Solution 5: Get Rid of the Roses

Pros: The beetles may leave and try to find food somewhere else.
Cons: Another predator might move in to eat the new plants. Also, there won't be any pretty roses in your garden anymore.

What did you decide to do? As you have learned, the stability of an environment depends on many things. Upset the balance and there's trouble in paradise!

a healthy rose garden

In this book, you learned about how plants and animals interact. You also learned about what humans do to environments. Ecosystems can be stable or unstable, just like a house of cards.

Take a look at the area around you. Is there something that threatens the stability of a habitat nearby? Is there pollution? Are there invasive plants or animals? Invasive species come from somewhere else, but they take over an area, like the Japanese beetles. You might be able to help with these and other problems. Start with research. Read books about the issue. Search the Internet for information. Talk to someone at a local environmental organization. Work with other people. Together, you can make a difference.

Purple loosestrife is an invasive species that crowds out other plants.

acid rain: rain or snow that is unusually acidic. Caused by air pollution, it can get into lakes or water supplies, harming wildlife.

adapt: to make suitable by changing or adjusting.

adaptations: characteristics that help an organism survive in an environment.

algae: plantlike living things that grow in water but do not have true roots, stems, and leaves.

altitudes: the heights of objects, such as mountains, above sea level.

condenses: changes from a gas to a liquid.

consumers: organisms that cannot make their own food, so they eat other organisms.

decomposers: organisms that eat and break down dead plants and animals.

ecosystem: a system of all living things, the nonliving things around it, and the relationship between the living and nonliving things.

elements: substances that cannot be broken down into other substances.

elevations: the heights to which objects, such as mountains, rise.

environment: everything around an organism that has an influence on it.

estuaries: the wide lower parts of a river where it meets the sea and the fresh water and salt water mix.

evaporates: changes from a liquid to a gas without boiling.

extinct: no longer existing.

habitats: places where an organism or a population usually lives.

hibernate: to sleep through all or part of the winter.

migrate: to move from one place to another seasonally.

photosynthesis: the process by which plants use carbon dioxide, water, and sunlight to make their own food.

population: group of animals of the same species.

predators: animals that hunt other animals for food.

prey: an animal that is hunted for food.

producers: organisms that make their own food.

scavengers: animals that eat dead animals.

species: a group of organisms that can reproduce with each other.

stability: the state of being firm and steady.

succession: natural changes in an ecosystem over time.

FURTHER READING

Bruchac, Joseph. *Rachel Carson: Preserving a Sense of Wonder* Golden, Colorado: Fulcrum Publishing, 2004.

Walliser, Jessica. *Good Bug, Bad Bug: Who's Who, What They Do, and How to Manage Them Organically (All You Need to Know about the Insects in Your Garden)* Pittsburgh, Pennsylvania: St. Lynn's Press, 2008.

The Franklin Institute Resources for Science Learning. **Living Things: Habitats & Ecosystems.** http://www.fi.edu/tfi/units/life/habitat/habitat.html

ADDITIONAL NOTES

The page references below provide answers to questions asked throughout the book. Questions whose answers will vary are not addressed.

Page 16: If the soil in a biome changed, plants might sicken or die and the animals that ate them or sheltered in them would have to move to a different environment.

Page 19: Decomposers like worms. They maintain stability of an environment because they tunnel through soil, adding air to it. They also turn dead plant matter into nutrients that plants can use. Without decomposers like worms, the soil would run out of nutrients to feed plants.

Page 21: Frog is predator and prey. It chases flies. It hops away from raccoons. Predator.

Page 24: From decade to decade, when the lynx population increases, the rabbit population decreases. When the rabbit population decreases, the lynx population decreases.

Page 25: The lynx population might decrease. The number of rabbits would decrease. Then the number of lynxes might also decrease.

Page 30: The pesticides could run off the garden into a stream or river that would eventually flow to the ocean. If it didn't rain for a long time, the pesticides would stay in the soil and affect plants and animals living there.

Page 31: During photosynthesis, plants use their leaves to produce food. Without the leaves, the rose plants did not get the nutrients they needed.

Page 32: If the climate changed, the plants that needed shade and lots of rain would die off. Other plants that thrived in these conditions would move in and replace them.

Page 33: They probably moved to a more suitable area. They or others like them will come back.

Page 37: The risks include poisoning helpful insects, poisoning animals such as birds that eat the insects, and polluting the soil and water.

Page 38: Japanese beetles are pests that kill plants. However, if there were no Japanese beetles, perhaps some plants would multiply too quickly.

Page 41: Fertilizers kill helpful animals and pollute the water and soil. You could use worms or other natural ingredients, such as compost, to enrich the soil.

INDEX

adaptations, 26

biome, 13–16

carbon dioxide, 31

Carson, Rachel, 35–37

climate, 10, 11, 13, 16

consumer, 18

cycle, 19, 29–31

decomposer, 19

ecosystem, 12, 13, 17, 23–24, 27, 28, 32, 38, 39, 40, 44

environment, 4, 8, 10-13, 14, 31, 34, 35–38

equilibrium, 23

fertilizer, 40–41

food chain, 22

food web, 22

habitat, 12–13, 17, 25, 26, 44

honeybees, 27

Japanese beetle, 6-7, 12, 18, 21–23, 25, 27, 37, 38, 42–44

nitrogen, 19

oxygen, 14, 31, 40

permafrost, 16

pesticides, 7, 30, 35–37

DDT, 35–36

photosynthesis, 18, 31

pollution, 23, 28, 39, 41, 44

population, 12, 24, 26, 27

predator, 7, 20–22, 24–25, 42

prey, 20–22, 25

producer, 18

scavenger, 19

species, 12, 34, 44

stress, 23

succession, 32

water cycle, 29–30